MY POLYCYSTIC OVARY AND ME

This Book is dedicated to

Abigail, mum Eileen and dad John

Who gave us the idea to focus this Book around a case-based format

We would like to thank our dear patients who volunteered their own stories to be added to this book. It was a generous contribution towards patient education about polycystic ovary syndrome.

The Endocrine Series

The Endocrine Series is designed to help educate you, the patient, to be informed of how to best look after yourself.

The first book "My Thyroid & Me", gives you detailed education about thyroid conditions and how to control them, monitor them and improve your quality of life. This book is Volume II and focuses on polycystic ovary syndrome. We are developing this series to teach you endocrine disorders.

The Endocrine series is designed to help you understand your own endocrine system to work better with your endocrinologist locally. It is a joint venture between you and your endocrinologist to try and achieve the best possible treatment outcome. By reading this series you will find out options of treatment available to you that you can discuss in depth with your endocrinologist. The information available on the net is often confusing and not tailored to your situation. This series of books is designed to help you learn about your condition, to be an expert patient and to help your endocrinologist look after you.

The endocrine glands are complex and often need fine tuning to work better. Have you ever thought why are you tired and why do you feel this way? Did you often wonder "I am just 50 why do I feel as if I am 90 years of age?" Did you look around you and ask, "how can a 70-year-old have more energy than me?" Hormones are responsible for all these well observed changes that you often cannot explain. An endocrinologist will listen to you very carefully and help you understand how your hormones regulate your body function to help you live each day better, excel and achieve your full potential. Most endocrine conditions are managed well and help you continue to enjoy your life.

Please note that this book is aimed to provide your education and knowledge about your condition however, it is not to replace expert opinions of your health care professionals.

The Cover

The cover represents the importance of exercise in the management of Polycystic ovary syndrome, the effect of adequate management in restoring the radiance of skin and the improvement of pregnancy outcome.

The Authors

Toqa A. Abdelrahman BPharm MClinPharm BCPS DIS
Assistant Lecturer of Pharmaceutics and Clinical Pharmacy
Mansoura University, Egypt

Toqa has worked in university teaching for undergraduate and postgraduate students at Pharmacy School, Mansoura University, Egypt for 5 years now. She has a deep interest in clinical practice and research, and public health. She received her Board certification as a Pharmacotherapy specialist from the Board of Pharmacy Specialties, USA in 2014. She was then awarded a Masters of Clinical Pharmacy from the Supreme Council of Universities, Egypt. She was trained as a Drug Information Specialist at the General Pharmaceutical Syndicate, Egypt. Later on she pursued further training in clinical research through joining a reputable clinical research training program by Harvard Medical School in collaboration with Dubai Harvard Foundation of Medical Research. Toqa aspires to develop a prominent career as a clinical researcher and practitioner, all with the aim of having more answers to medical dilemmas to offer a helping hand to patients all over the world.

Dr Hisham Maksoud MD FACE FRCP Edin FRCP London LLM
Consultant Endocrinologist UK

Hisham is a consultant Endocrinologist who runs busy Endocrine clinics in Nottingham UK. He has special interest in teaching. Trained at the Oxford Teaching Hospitals. He received his training in Endocrinology based on clinical research and spent five years in research as part of his Endocrinology training. Hisham travels regularly to USA to learn from the latest research that he can apply to his clinic. Hisham is a member of the Endocrine society in the UK and USA. He is a fellow of the American association of clinical endocrinologist, Royal College of Physician Edinburgh, and London. His research is focused on patient education and how to translate current research to clinical practice today.

TABLE OF CONTENTS

SECTION 1

DIAGNOSIS

Why do you need this book?

Many patients attend to the clinic with a significant number of unanswered questions regarding polycystic ovary syndrome. The patients often are not aware of the presentation of polycystic ovary syndrome and there has been a significant amount of confusion because of the published data on the internet. Some websites are outdated and the information that we have received from research over the last 10 years has explained in more detail the different clinical presentations of polycystic ovary syndrome that have been identified. There has been a need to have a clear concise up-to-date information book to explain to the patient in plain straightforward language what polycystic ovary syndrome is all about.

What is Polycystic ovary syndrome?

Polycystic ovary syndrome is the most common endocrine condition in a young lady from age 15 to 50 and does not stop by the onset of the menopause. In fact, its effect on the metabolic system becomes worse after the menopause due to the development of diabetes, high cholesterol and high blood pressure.

This important condition affects young females as well as older ladies and it manifests in many ways. Some patients have excessive hair growth, some patients do not, some patients are overweight, and others are not. However, all patients have the common issue of increased insulin level and insulin resistance which results in increasing weight and difficulty in controlling the blood sugar unless the patient exercises aggressively to maintain their weight. That is why some patients having polycystic ovary syndrome do not present the usual picture we see in 50% of cases where patients tend to be overweight with excessive hair growth and acne, rather the other 50% of patients are slim.

Not only that, but some patients do not even present with acne or excessive hair growth and are unaware of the subtle abnormality noted on presentation. General practitioners often write to us and express surprise that the patient has polycystic ovary syndrome while having a regular period. Again, that is another misconception, as a patient with polycystic ovary syndrome can have a regular period. Patients with polycystic ovary syndrome may not have excessive hair growth or may not have acne as the condition can be masked by the oral contraceptive pill that the patient has been on for some time or the fact that the patient exercises excessively.

The purpose of this book is to answer all the above questions in a clear concise manner. This book is a manual to help you explore the current symptoms that you have and understand your own symptoms. This book will help you explain your condition to your health care professionals and to guide them through the difficult diagnostic dilemma of polycystic ovary syndrome.

If you are diagnosed with polycystic ovary syndrome, the condition is treated in a very straightforward manner, address clear checks that you need to follow to try and help you manage and avoid the complications caused by this condition.

Background of polycystic ovary syndrome

Polycystic ovary syndrome is a set of symptoms that develop due to raised testosterone-like hormones in the blood in ladies. Their insulin level rises and causes further problems with weight gain and sugar level disturbances. There are signs and symptoms of polycystic ovary syndrome. These could be the absence or irregularity or heaviness of the period and excessive hair growth and acne. Pelvic pain; that for some reason affects the left side more than the right, difficulty getting pregnant or difficulty losing weight or significant mood changes or sugar cravings are also common symptoms of the condition.

The symptoms can be triggered shortly after puberty and can develop during the later teenage years and early adulthood and becomes worse as we progress in life. The difficulty of course is making the diagnosis and often the patient presents quite late unless their mother has had the syndrome and aware of the different presentations.

We were lucky enough to meet one of our patients, RG who had a very clear idea about this syndrome which had affected her and brought her daughter early on to have a clear diagnosis. By the action of RG, her daughter EG started on the treatment early enough. It is not unusual for the patient to present as part of a family genetic signal that has resulted in the development of the condition.

Now it is also important to note that polycystic ovary syndrome is not a disease, it is a clinical syndrome which is a collection of symptoms and signs that present in a certain way. A diagnosis of polycystic ovary syndrome does not necessarily mean you will never have a baby of your own or that you are infertile. That is not true, in our clinic we seem to have very few patients who have problems with fertility if they are being treated properly for the polycystic ovary syndrome early enough. This very important point has been misrepresented on the internet and in several books.

Polycystic ovary syndrome is an important condition and it is important to understand its impact upon you. You need to learn how to regulate and manage the testosterone-like hormone effect on you and the high insulin level that is the whole marker of this condition. You can significantly improve your quality of life by simple means such as exercise and diet.

SUCCESS STORY 1

Patient background

KB is a delightful 38-year-old teacher. KB was referred by her GP for the management of an underactive thyroid. She was stable on her then current thyroid treatment for many years. However, three years prior to being referred KB struggled with an underactive thyroid and a raised thyroid stimulating hormone. The GP commented that she is very compliant with her thyroid. She denied clearly that she has missed any of her medications and was offended about possible allegations of non-compliance. She asked for an expert second opinion and hence the referral to our unit.

Referral to the endocrine unit

KB was referred to us in 2010 when she was 30 years old. We reviewed KB and listened very carefully to her symptoms. She noticed that the hair on the scalp was falling, she had thinning of the eyebrows and had developed acne on the forearm and the legs since age 18. She was unable to enjoy a long-term relationship because of the significant tiredness which meant that she could only work and sleep. She did not have enough energy to socialise. She did not have an ultrasound of the ovary before and she developed a condition called "ulcerative colitis" which is an inflammation of the colon. KB had been on the oral contraceptive pill since age 13 due to heavy periods. Her weight had increased suddenly from 9 stone at the age of 21 to 18 stone (115 kg) at the age of 30. She was upset by the fact that she was unable to lose weight despite regular exercise three times a day.

The key area here was that KB had more than just a thyroid issue but there is another hormonal problem in play that caused the current presentation including the thyroid abnormality. We suspected that KB had polycystic ovary syndrome causing her the difficulty losing weight, the acne and the hair loss. Therefore, we organized for KB an assessment of her steroid system and hormone profile in detail. Upon careful assessment, KB was noted to have her luteinizing hormone raised compared with her follicle stimulating hormone. We did not really need an ultrasound to diagnose polycystic ovary syndrome as KB fulfilled two out of three Rotterdam criteria which are: evidence of acne and irregular heavy periods; irrespective of what the ultrasound showed. Often the ultrasound tends to be misleading if the patient had been on oral contraceptive pill and KB was on oral contraceptive pill from age 13 to age 30. That had wiped out any evidence of cysts on the ovaries.

Turning point

Fast forward to 8 years down the line, KB has done remarkably well. KB worked extremely hard on managing the polycystic ovary syndrome with diet and exercise, Metformin and Spironolactone. KB also required further help to manage her insulin resistance with Victoza. This is a medication that has been used in research papers to help patients with polycystic ovary syndrome control their insulin resistance and patients with difficulty losing weight achieve an astonishing amount of weight loss.

We reviewed KB at the endocrine clinic in 2018 and the results are astonishing. KB's weight dropped from 18 stone (body mass index of 38) down to 13 stone 2 lbs. (body mass index of 28), and she felt so much better. The significant weight loss that KB has achieved has prevented the onset of type 2 diabetes. KB's energy has significantly improved, the sweating that caused KB to abandon social events after 20 minutes has improved, she exercises regularly, and she is now able, as a teacher, to take part in physical education classes.

Over the last three years KB attended to the endocrine clinic every three to four months for regular review. KB achieved all that by her determination and eight years of hard work on diet, exercise and improving her thyroid function and not just by taking medications. Her thyroid issue has now resolved, and she feels so much better.

How the diagnosis of polycystic ovary syndrome is made?

Polycystic ovary syndrome is a condition that has been defined in many different medical books and resources. It is a very common condition that received great attention from specialists worldwide. Hence, a list of diagnostic criteria was reached based on input from leading endocrinologists from Europe and USA who met in Rotterdam in 2003. These Rotterdam 2003 criteria are now the standard for making a diagnosis of polycystic ovary syndrome. A diagnosis of polycystic ovary syndrome is made if you have any two of the following three criteria:

Criterion 1: infrequent or absent or heavy periods (remember you must not be on the oral contraceptive pill for at least 6 months otherwise you will mask this criterion)

Criterion 2: Presence of cysts in the ovary

Criterion 3: Excessive hair growth or acne or biochemical evidence of a raised hormone signal called luteinizing hormone, which is a signal that is triggered by the pituitary gland in the brain and has an impact on the ovary to regulate the ovarian function with another hormone signal called the follicular stimulating hormone.

Therefore, the definition of polycystic ovary syndrome can vary depending on whether you have any two out of these three criteria. For example, the patient may not have cyst in the ovary yet have polycystic ovary syndrome. The diagnosis was confused by the name "polycystic ovary syndrome" as a patient may have the syndrome but without having cyst in the ovary.

The pituitary gland is an important part of the syndrome which we will discuss in the later part of this book. The luteinizing hormone is triggered by the pituitary gland signaling for the follicular stimulating hormone to regulate the ovarian function. That is where all the syndrome starts, in the brain not in the ovary. The ovaries are accused of something they have never done. They are simply responding to a request from your pituitary gland. Please be careful and read carefully any internet articles as they seem not to get this point correct.

So, do you have polycystic ovary syndrome?

If you were to be diagnosed with polycystic ovary syndrome, you need to explore whether you have any two out of the three criteria that we have highlighted above which are: irregularities of the period; you must not have had oral contraceptive pills for 3 years for this symptom to show as a presentation of the condition as oral contraceptives tend to control the effect of the syndrome on you and often lead to misdiagnosis as we explain later in the book. The second important criteria are whether you have acne or excessive hair growth, and where do you look for the hair growth.

Excessive hair growth is mainly on the chest, the abdomen, the back of the leg, the toes or you may not have any, but the facial hair can be easily spotted on the jaw line or the mandibular area, at the chin and over the upper back. Yet, we still have patients with polycystic ovary syndrome who do not have acne and do not have hirsutism (excessive hair growth). As you can see it is one criterion out of the three and you only need two criteria to be diagnosed with polycystic ovary syndrome.

Therefore, the diagnosis for polycystic ovary syndrome is not straightforward as it sounds. It requires careful meticulous history-taking and exploring your current treatment to be able to make a diagnosis. If you are suspecting that you have polycystic ovary syndrome you should ask your general practitioner to refer you to an expert endocrinologist with a special interest in this field. It is important to be off the oral contraceptive pill for a minimum of 6 months prior to having a blood test for diagnosing polycystic ovary syndrome. Please be careful to consider, that if you are on the oral contraceptive pill for the last 3 years prior to the diagnosis that may affect your results.

What other conditions must be ruled out if I'm diagnosed with polycystic ovary syndrome?

You cannot be diagnosed with polycystic ovary syndrome without clear assessment of specific areas in your endocrine system.

Cushing's Syndrome

Cushing's syndrome is a condition where there is excessive cortisol secretion, and this can mimic polycystic ovary syndrome. The best way to assess whether you have Cushing's Syndrome or not is by a simple blood test that your GP can perform and that is a cortisol level test having had a tablet of Dexamethasone the night before. If your cortisol level is recorded as under 50 then you do not have Cushing's syndrome in most cases. There are rare conditions where this test fails and if that happens you can then have a review by an endocrinologist to explore the other rare presentations of Cushing's syndrome.

High Prolactin level

Another misdiagnosis is patients who have high prolactin who are diagnosed with polycystic ovary syndrome. A raised prolactin can also stop your period even if you do not have polycystic ovary syndrome. Raised prolactin can be a stand-alone abnormality that is not necessarily a part of polycystic ovary syndrome. Therefore, it is important to correct your prolactin first before you are diagnosed with polycystic ovary syndrome. The treatment for a raised prolactin is very straightforward and it depends on several factors such as the medications you are on and whether you have any disturbance of the pituitary. An assessment of the pituitary gland can be very straightforward, and your endocrinologist can help you with that.

Thyroid abnormalities

A disturbance to your thyroid hormones can also mimic polycystic ovary syndrome as we mentioned earlier. A heavy period can be caused by underactive thyroid and a period can stop because of overactive thyroid. It is important to realise that the thyroid can always disturb your period, can mimic a polycystic ovary syndrome picture and can upset your fertility.

Stress

Stress itself can upset your period and can mimic a polycystic ovary syndrome picture. It is important that you manage your stress level better as the stress can cause you an increased cortisol level, increased insulin level and disturbance to your period. An increased cortisol level induces excessive hair growth, disturbance to your quality of hair and hair loss on the scalp.

Once you have cleared all these important issues then, the diagnosis of polycystic ovary syndrome is straightforward by identifying two out of the three Rotterdam criteria which are: development of acne, excessive hair growth, irregularity/ disturbance or absence of the period, and cyst of the ovary.

If you were diagnosed with polycystic ovary syndrome, you must start working very quickly to try and control your weight if you are overweight or control your insulin resistance if you have a normal weight. This will help you avoid complications such as the development of diabetes, high cholesterol and endometrial cancer (womb cancer). The weight management is very straightforward and requires you to address the diet and exercise as detailed below in the treatment section.

You have to avoid the onset of diabetes as it can develop in up to 40 % of ladies with polycystic ovary syndrome if they are not treated. If you are treated, then your risk of developing diabetes diminishes significantly. It has been advocated in the literature that adopting a high fat diet will help you lose weight as a high fat diet can exaggerate the already disturbed metabolism and cause you high

SECTION 2

TESTS

Why have a blood test for polycystic ovary syndrome?

If we can make a diagnosis of polycystic ovary syndrome clinically, by means of asking you if the period is on time, looking for history of acne or hirsutism, then why do we order blood tests?

The reason we do the blood tests and the ultrasound scan is to look for other conditions that could mimic polycystic ovary syndrome. For example, patients with high prolactin can have problems with the period and the period can stop, patients with thyroid problems can have problems with heavy periods or absence of the period.

The most difficult part of course, is that the problem with your thyroid can affect your weight, can affect your metabolism and can cause you problems that are like a polycystic ovary syndrome presentation. Patients who present with Cushing's syndrome (excessive cortisol levels) tend to have similar presentation to polycystic ovary syndrome patients and testing for that is very clear as we explain in the test section later in the book.

So, the key here is that polycystic ovary syndrome is a condition that must be diagnosed very carefully. The blood test can help you define whether there is a disturbance to the thyroid, the prolactin level, or your cortisol levels.

SUCCESS STORY 2

Patient background

DR is a 51-year-old prison office. She was diagnosed with type 2 diabetes in June 2014. She weighed 116 kg at the time and was having extreme difficulty losing weight with her current life style and work commitments. DR struggled with a persistent high level of blood sugar and a raised blood pressure of 155/95. The GP started DR on Orlistat, which is a medication that minimises fat absorption from food in the gut and hence is used as a weight loss medication. DR however developed disturbing side effects to this medication including headaches and diarrhea. The GP discussed bariatric surgery with DR as an option to help her lose weight, however she was very reluctant to do so.

DR tried to lose weight on her own without success and the GP then started Metformin to try and take control over the blood sugar and the insulin resistance. DR was again very intolerant to the side effects of the medication which included diarrhea and an abnormality in liver function tests detected on further assessments.

The GP next tried several oral hypoglycemic drugs (medications that lower blood sugar) however her high blood sugar persisted, and she was now 118 kg in weight (equivalent to a body mass index of 43). DR developed side effects to the blood sugar lowering medications including dry eyes and mouth and all medications were withdrawn at this point. The GP was running out of options and referred DR to the endocrine clinic in July 2015.

Referral to the Endocrine Unit

DR attended to the Endocrine Clinic and was reviewed carefully. She explained her continued struggle with obesity despite her efforts to bring the weight down. She was frustrated, fed up and really needed help. We listened very carefully to DR and upon review she was noted to have insulin resistance, water retention, hirsutism, weight gain with a weight of 18 stone 7 lbs improving to 18 stone after repeated trials of medications.

DR had an irregularity of the period, excessive hair growth and hair loss over the scalp. Her hormone profile showed a raised luteinizing hormone to follicle stimulating hormone ratio which is again characteristic of polycystic ovary syndrome.

It was by now evident, that DR's diabetes may not actually be the main problem in this complex clinical scenario. DR now at the age of 51 is having a late diagnosis of polycystic ovary syndrome that may have been complicated by the diabetic picture she is having now. The diagnosis was not out of question and an ultrasound was not required for confirmation since DR already had two out of three of the diagnostic criteria described earlier.

This poor lady had developed a disrupted sleep pattern due to working night shifts frequently and a drop-in oxygen level during sleep causing her to wake up in the morning with headaches. Furthermore, DR was noted to have frequent urinary tract infections that can be as many as five episodes per year despite multiple antibiotic courses and had recently struggled with a prolonged chest infection. This indicated disruption of her immune system which is a common issue with untreated diabetes and persistently high blood sugar level.

Turning point

An action plan was developed to aim for specific milestones in the treatment of this delightful lady to help her resolve the condition and to take things into control. It was very important to eradicate DR's recurrent urinary tract infection. To do so, she was started on an antibiotic course and her low Vitamin D level was addressed using adequate Vitamin D supplementation.

DR was started on a comprehensive diet and exercise program to try and manage the insulin resistance. DR was also started on Victoza to help her with the weight loss and Spironolactone to help her with the hair loss over the scalp and excessive hair growth on the body.

DR worked very hard on this treatment plan and we contacted the occupational health service to try and limit the night shifts she worked on, as this was causing her an abnormal cortisol level which further affected the insulin resistance.

In November 2015, DR had started to improve and by July 2017, her blood sugar was improving, her Vitamin D improved, she has regained her work-life balance and lost 2 stone in weight, she is no longer having recurrent infections and the reason for that is avoiding the night shifts together with the diet, exercise and Victoza. By September 2017, DR had lost a further half stone in weight and was feeling much better.

Vitamin D is vital for your immune system to function properly and it is very important to make sure that your Vitamin D level is optimized especially in such a case when the immune system starts to fall back behind the body's ability to fight infections. A healthy sleep pattern can significantly alleviate the stress posed on your body which can cause a disturbance in your cortisol level.

Cortisol level greatly affects the ability of your body to function and worsens insulin resistance and therefore complicates diabetes and polycystic ovary further. It is also very important to understand that many endocrine issues show interlacing symptoms that can lead to an important diagnosis being overlooked.

DR did have diabetes however that was only a complication of the polycystic ovary syndrome and once the correct diagnosis was made and managed, the patient started to notice significant improvement in the quality of her life and health, the diabetes was reversing, she was on target to lose the excess weight and felt so much better.

What tests do I need to diagnose polycystic ovary syndrome?

We often look for evidence of raised luteinizing hormone compared with a level of follicular stimulating hormone and tend to repeat the test usually on three occasions, each a week apart. We also look for evidence of thyroid abnormality using thyroid stimulating hormone and free T4.

We test for a disturbance in androgen-like hormones (testosterone-like hormones) by testing levels of testosterone, dihydroepiandrosterone and androstenedione. We look for evidence of Cushing's syndrome by testing the cortisol level. Finally, we order an ultrasound to make sure there is no ovarian tumor or abnormality within the pelvis. In addition to that of course we are looking for cysts in the ovary. However please remember that you may not have cyst in the ovary even if you have polycystic ovary syndrome and if you don't have polycystic ovary syndrome you may still have cysts in the ovaries.

Therefore, the existence of cysts in the ovary does not in itself define a diagnosis of polycystic ovary syndrome. The tests for polycystic ovary syndrome are, as we have explained earlier, confirmatory rather than diagnostic. This means these tests will help define whether you have other conditions that can mimic polycystic ovary syndrome but does not necessarily make the diagnosis.

The diagnosis is made clinically by your physician and by you listening very carefully to your body if you have a problem with any of the issues highlighted above. It is straightforward, and you can use the diagnostic criteria to guide you towards your diagnosis if you have two out of the three criteria mentioned earlier in this manual.

Misdiagnoses of polycystic ovary syndrome

Misdiagnosis of polycystic ovary syndrome can happen if the patient is slim. 50% of patients with polycystic ovary syndrome are slim and the reason is that they have a certain exercise program that they have developed to maintain their weight. By maintaining the weight under control, they then tend to avoid the evidence of developing the symptoms of polycystic ovary syndrome. If the patient develops an injury or is not able to exercise for any reason, they then struggle with excessive hair growth, development of acne or increase in weight.

The second important misdiagnosis is the misconception that the patient must be overweight to have a diagnosis of polycystic ovary syndrome. Not at all, this is not the case in 50% of cases we have patients again as we have highlighted above not overweight who can have polycystic ovary syndrome.

The third important misdiagnosis is ruling out polycystic ovary syndrome when the patient does not have acne or excessive hair growth. This is not true, as we have patients without acne or excessive hair growth and still have polycystic ovary syndrome. The hair growth can be very subtle, it could be just around the nipple area, the abdominal area, the toes or the back of the leg.

The patients sometimes accept the fact that the hair growth that they have is a part of their racial profiling. Some cultures have a certain hair growth pattern that is expected as a norm when in fact it is a sign of polycystic ovary syndrome.

Patients can present with hair loss over the scalp and the patient overlooks this sign of polycystic ovary syndrome because they do not have any other features to suggest polycystic ovary syndrome. Hair loss on the scalp is an important feature of excessive androgen (testosterone-like hormone) effect of polycystic ovary syndrome on the patient.

Another reason the diagnosis is missed that the patient does not have any feature of polycystic ovary syndrome because they have been on the oral contraceptive pill for many years. We have patients on oral contraceptive pills for as long as 20 years and of course, by the time they come off the pill to try and become pregnant they struggle with excessive acne or excessive hair growth and they struggle to explain these important symptoms having not had problems with the period, acne or hair growth for 20 years. Of course, what has been happening is that the pill has been masking the polycystic ovary syndrome picture.

JE is one of our patients who presented lately with features that suggest polycystic ovary syndrome such as increased hair growth, increased acne and increase in weight and has managed extremely well to control these important symptoms having not had any of these symptoms for 20 years. She stopped the oral contraceptive pill and started having her children. By the birth of her third child, the symptoms started to develop due to the triggering of the background of polycystic ovary syndrome that has become much more obvious with mood change, excessive hair growth and acne.

If in doubt about the diagnosis, what do you do?

Ask your GP for a referral to an endocrinologist. Read and record your symptoms over a three months period. Write down on paper how your symptoms affect you on a weekly basis. Try and time your period and your meal time to your symptoms.

SECTION 3

THE PITUITARY GLAND

Why the pituitary gland is important in the development of polycystic ovary syndrome?

The pituitary gland is a very small gland, less than half a centimeter located in the middle of the head and manages all your hormone profile. It controls your ovary, thyroid and adrenal glands, your growth hormone, water balance and milk production at the time of breast feeding. It is where the luteinizing hormone and follicle stimulating hormone are released to control the ovarian function. The disturbance to the luteinizing hormone and the follicle stimulating hormone is what triggers the symptoms and signs of polycystic ovary syndrome and it is also what triggers a disturbance to the insulin level.

Copeptins are proteins made by a gene called KISS1. They were first discovered in 1996, in Hershey, PA, USA and were subsequently named after Hershey's chocolate kisses. The Kisspeptins seem to control the luteinizing hormone and follicle stimulating hormone at the pituitary gland and can be a trigger for polycystic ovary syndrome. Hershey's Kisses is chocolate made by The Hershey Company. It has a distinctive shape as a teardrop and it is wrapped in squares of lightweight aluminum foil pointing to the top.

"You're sweet like chocolate

Finding your way in the dark

Ain't so hard when you're close to my heart

You are there when I'm feeling alone

All I need is for you to come home

--Lady Bee - Sweet Like Chocolate-- "

Therefore, by looking after your pituitary, by careful rest, adjusting your calorie intake, ensuring that you have less stress during your day and ensuring an adequate sleep pattern are crucial. These measures could help the pituitary gland to better manage the disturbance that develops in the luteinizing hormone and the follicle stimulating hormone and the rise in the insulin level to help you control the period and the weight better. That is why, when ladies develop anorexia, their period stops. Simply, because the disturbance to the weight upsets the pituitary and the pituitary decides to shut down the period. In a similar way, when patients have a disturbance to their pituitary gland with a raised luteinizing hormone and follicle stimulating hormone levels, that results in a disturbance to their period, problems with managing how their female hormones function and the resulting weight gain.

We also know that a raised prolactin hormone from the pituitary gland can develop in patients with polycystic ovary syndrome and can result in a disturbance to their weight and period, and by controlling the prolactin the period is restored.

The Problem is Insulin Resistance

What is insulin resistance?

Insulin is the hormone that is responsible for opening the cell gate to allow the sugar in and that is why some patients with diabetes who do not have insulin or enough insulin have high blood sugar. This is because the insulin that is required to open the cells to let the sugar in is absent or deficient.

Insulin resistance is a major issue that develops in polycystic ovary syndrome. It is a whole marker of this important syndrome. As you get a disturbance of the luteinizing hormone and follicle stimulating hormone, your insulin level is disturbed as well, and this presents as weight gain, fluid retention, sugar craving and sleep disturbance. You wake up at 3 in the morning with a restless sleep, you start to snore at night due to sleep apnea and you start to develop high blood pressure and diabetes if this is not treated.

Insulin resistance may present as a genetically related condition, it can be acquired if you have been on steroid for a long time and it can develop because of other rare metabolic conditions. However, it is a common issue that develops in polycystic ovary syndrome.

Your cells become resistant to insulin and your pituitary starts signaling for more insulin production to force cells to take in the blood sugar. Because of this significant rise in the insulin level, an increased risk of diabetes develops, you start struggling with increased sugar craving especially after the evening meal (your cells are not taking in the blood sugar) and you have restless sleep due to a drop in the blood sugar in the middle of the night that causes you to wake up because the brain does not have enough sugar to function at night.

An increased insulin level causes weight gain because insulin is a growth hormone, it makes patients grow and that is why patients with polycystic ovary syndrome develop weight gain that is difficult to lose. Therefore it is key to address the insulin resistance first with diet and exercise, to start achieving your target weight.

SECTION 4

TREATMENT

What treatment is available for polycystic ovary syndrome?

Why treat polycystic ovary syndrome?

Polycystic ovary syndrome is a very important condition. It can cause type 2 diabetes, raised cholesterol, disturbance to the lining of the womb and difficulty losing weight. For patients who are slim and don't have an issue with the weight, exercise is important and if they were to stop the exercise the weight will increase and cause further problems. In addition to that, patients who are slim can have problems with hair loss over the scalp, acne and excessive hair growth and that is why it is important not just for the patients who are overweight but also for the patients who are of normal weight.

That is why we need to treat polycystic ovary syndrome patients. It is important that you have a clear idea of what is the effective treatment. The best treatment for polycystic ovary syndrome is diet and exercise. However, it is difficult to implement that on daily basis due to our lifestyle, due to the structure of the day, the pressure of work and the disturbance to the sleep pattern.

SUCCESS STORY 3

Patient background

LB is a delightful 25 year old who was referred in 2013 with a history of painful heavy periods. LB complained of weight gain especially around the abdominal area and felt quite emotional. LB felt quite tender in the right side of the pelvis. LB complained "I don't feel right". Her GP informed her that the blood test and the ultrasound do not suggest polycystic ovary syndrome. The GP asked for an expert second opinion.

Referral to the endocrine unit

LB was then reviewed in our clinic. We listened to LB and reviewed her history carefully and it was quite interesting to note that she has polycystic ovary syndrome. The key area in the history is the fact that she has acne and erratic periods. LB's blood tests indicated that her luteinizing hormone was much higher than her follicle stimulating hormone. She also struggled with weight gain, was quite anxious and low in mood. She has a history of irritable bowel syndrome.

We discussed with LB the diagnosis of polycystic ovary syndrome and explained the guidelines for the diagnosis as clearly defined by the Rotterdam criteria as having two out of three of the following criteria: 1 erratic period, 2 raised luteinizing hormone to follicle stimulating hormone ratio or evidence of acne or excessive hair growth and 3 evidences of cyst in the ovary. The difficulty in that of course was that LB was on the oral contraceptive pill because of her erratic period and therefore her ultrasound was clear of any cysts on the ovaries.

Turning point

LB was then started on Metformin to reduce the insulin resistance and was started on Spironolactone to help her control the acne and the hair loss and was started on the oral contraceptive pill to help her manage the irregularity of the period. When LB first presented her luteinizing hormone was 74 and her follicle stimulating hormone was 9.

In 2017, LB lost a significant amount of weight, she felt so much better and continued the treatment. We encouraged LB to try and come off the Spironolactone however she was very reluctant due to the significant improvement in her skin and her hair. The acne significantly improved as well as the irregularity of the period and she felt so much better. The hair loss continued to be an issue in certain times of the year and we have agreed on providing LB with further support to help her manage the hair loss better.

LB's diet was very strict in terms of avoiding high carbohydrate diet and ensuring she has food with a low glycaemic index. This meant that she was to avoid potato, bread, pasta, water melon, pineapple, mango, grapes, dates and banana. In addition, her Vitamin D was low, and she was started on Vitamin D and she felt much better.

LB continued to improve and we have agreed to monitor the treatment and LB has regular blood tests to make sure that she continues to have clear evidence of safety of the current treatment. Her thyroid is being monitored carefully to make sure that she does not develop hypothyroidism.

Reflection

If the patient was on oral contraceptive pill, the period will regulate, and that part of the history is dismissed. That is often the issue with the diagnosis of polycystic ovary syndrome as the use of oral contraceptive pill tends to mask out evidence of cysts in the ovary making an ultrasound look abnormal when in fact it is just being masked by the medication. The presence of cysts on the ovaries can help the diagnosis but does not necessarily mean that every patient with a cyst on the ovary have polycystic ovary syndrome.

Therefore, it is important that when you speak to your physician you make sure that you give them a clear timeline of when and how often you used the oral contraceptive pill. It is important to outline the use of the pill from the onset of your period (menarche) to your current age and indicate clearly for how long and when did you use the oral contraceptive. This can provide an explanation for the period being well controlled by the oral contraceptive pill and when that is stopped, the period can still be regular for a year or two before it becomes irregular once more.

Another key area to point out here, is that the thyroid function should always be monitored closely in polycystic ovary syndrome patients to keep an eye on possible hypothyroidism. According to current research, 20 % to 40% of polycystic ovary syndrome patients tend to develop underactive thyroid disease and hence the close monitoring helps catch that early on before the patient starts having symptoms.

There are specific medications that a patient can consider for hair loss such as Minoxidil or Vitamin supplements such as Viviscal. However, we do not really recommend them to start with. One should not resort to these treatment unless all recommended management of the polycystic ovary syndrome have been already undertaken and optimized.

Instead, we first recommend that you address the weight with diet and exercise and then consider introducing the medications that address the insulin resistance and the high testosterone drive that cause the acne and the hair loss, before moving on to use further alternatives as the ones mentioned above.

Avoid laser therapy until you have addressed the basic treatment of PCOS for at least six months. What that does is improve your treatment success with laser therapy by at least 60 %. It is so sad to see a number of patients in the clinic who spent significant amount of money on laser therapy and did not address the basic treatment of PCOS first.

Diet Plan

There has to be a clear diet plan which can be constructed carefully by examining your daily dietary intake and adjusting the diet very little to result in significant improvement in your quality of life.

We would recommend you avoid the following nine types of food: Potato, bread, pasta, watermelon, pineapple, banana, dates, grapes and mango.

Why specifically these types of food? These foods tend to have high glycaemic index and they result in an increased weight significantly. Other types of foods that are available to you should not cause you any further problems if you were to avoid the above-mentioned types of food.

It is important to note that it would be a good idea to stay away from Gluten-based diet. It is important to limit the portion of each meal and to ensure that you have a small breakfast, small substantial lunch and small dinner. It is important to try and move your dinner time to an earlier time of the day rather than later.

You may switch lunch with dinner to make sure that lunch is much more of a substantive meal for your day. Allow the dinner to be the smallest as by the time you have dinner, the day is over, you will have very little time to exercise and you tend to be very tired after a long working day. This will mean that the food will tend to accumulate around the abdominal area because the insulin did not have time to move it forward to the next part of the metabolic system to utilize it.

For example, I have a very good patient who is called RB, she works in Leicester and she finds it very difficult to find time after a 12-hours working day to have her meal early. Therefore, we agreed that on the days that RB arrives late from work, she can only have salads and then retire to bed. Other days when RB is earlier at home she can have the evening meal as more substantial by 5 or 6 o'clock in the evening with her partner. A simple adjustment like that can make a big difference to the way that your body can manage the food load that you ingest.

It is also important to try to think of ways to reduce the workload on the pancreas and the insulin requirement. The way to do that is to think of two days of every week in which you restrict your food to greens and fruits. Of course, this does not include breakfast, but you should prefer salads for lunch and the evening meal. Many cultures practice fasting for religious reasons and a similar weekly regime that allows for two days of green diet makes a significant improvement to your blood sugar and the cholesterol profile.

As you may have noticed, we have not mentioned any medical treatment for polycystic ovary syndrome so far. Medical treatment is only a part of the management of this important condition whereas the bulk of the treatment is diet and exercise.

Exercise Plan

The exercise must be simple, it has to be practical, it has to fit into your working day and it does not have to burden you on top of a very busy schedule. The best exercise regime is to work out three times a day for 10 minutes each time. For example, why not get up 10 minutes earlier and have an exercise bike close by? Work out in some place that is convenient for you; the sitting room or the bedroom perhaps. Spend 10 minutes only, not more than that, in the morning, right after you arrive from work and last thing at night. You do not have to book time to dress up and go to the gym, it is a part of your day.

It is also important to think about a 2-minutes' walk booked into your daily schedule. If you were to do that, you are going to do at least 30 to 50 minutes of exercise time daily building up to around 350 minutes a week and that is all you need to reduce the amount of insulin associated with polycystic ovary syndrome that causes you difficulty losing weight. 210 minutes per week for an exercise of 30 minutes per day or 350 minutes a week if you incorporate a 20 minutes' walk to a total of 50 minutes a day.

You can leave your home for a quick 10 minutes' walk and then come back that is how you can add 20 minutes' walk per day into your busy day. Think of it how long do you sit there in traffic Jams. Can you imagine if you use this time to simply walk. Walking is so powerful. It stimulates your heart and pump oxygen around all your cells and reduce the risk of dementia.

Every 66 seconds someone in the USA is diagnosed with dementia. Can you imagine what will happen if we get everyone walking 20 minutes a day. Health is so important and easy to achieve if we book it into our diary if you do not sit down and book it in your diary it will never happen. When we have to travel abroad for research and do not book time to exercise we never do it.

Research performed in many laboratories around the country and in the USA emphasized how important it is that a simple exercise program can result in significant weight loss. The key area here is that diet and exercise are the most important component of the management of polycystic ovary syndrome.

Other Treatment Options

Of course, there are other measures to manage polycystic ovary syndrome and that include oral contraceptive pill, Metformin, Spironolactone.

Oral Contraceptive Pills

We know that oral contraceptive pills can improve the hair growth and the acne however it can also disturb the weight. For example, if a patient has a very heavy or disturbed or painful period or excessive hair growth or acne, the oral contraceptive pill is a very useful way of managing these difficulties. However, it does upset the weight in some patients and this has been published clearly in many papers.

Therefore, if you were to use the oral contraceptive pill for contraception purposes or for managing the excessive hair growth or the acne or the irregularity of the period, you need to be aware of the limitations and the impact of the pill positively on your period, hair and acne but also negatively on your weight. This means that you will need to address your weight with diet and exercise much more aggressively than any other patient.

SUCCESS STORY 4

Patient background

TP is a delightful 20-year-old who was suspected to have polycystic ovary syndrome. She presented with acne and heavy periods.TP was started on oral contraceptive pill for the management of her symptoms however she was intolerant to the pill despite trying Marvelon, Microgynon and Yasmin. She had developed unpleasant skin changes as a side effect.

The oral contraceptive pill was discontinued twelve months later by her general practitioner. She asked for an expert opinion in view of her current symptoms. Her luteinizing hormone was raised at 17.8 and the follicle stimulating hormone was 6.1, her testosterone was raised at 4.9. The normal testosterone for a lady is under 1.5. The period stopped 18 months ago.

Referral to the endocrine unit

TP was referred by her GP two years ago. We reviewed TP at the endocrine clinic quickly within a few days of her referral and her symptoms were significant enough to cause concern. She was struggling with lethargy, difficulty losing weight and she felt very tired. She had a number of treatments for acne, she was even started on Roaccutane. Her hair was greasy and had excessive harsh hair growth on the body specially the abdomen, the chest and the legs. TP struggled with sugar craving. She consumed a significant amount of carbohydrates to be able to function daily. We discussed polycystic ovary syndrome and we started TP on Spironolactone, Metformin, Vitamin D and Vitamin B12 in addition to Dianette. We agreed on a diet and exercise program.

Turning point

One year ahead, TP had improved to a significant degree, her hair was excellent, the quality of her skin significantly improved, her period was on time (regular) and she felt so much better. Her Vitamin D was improving to an excellent level.

We reviewed TP again a few months ago and she continued to improve with excellent results. She achieved a total normalisation of her testosterone and luteinizing hormone and her quality of skin, hair and mood have become "fantastic". TP continued the treatment with Spironolactone and Dinette and did not need the Metformin any further.

Therefore, this is an example of a lady who improved significantly just on Spironolactone and the oral contraceptive pill. Eventually, TP did not need Metformin as part of her treatment as her sugar craving improved significantly, in addition to having had some side effects to the Metformin.

Reflection

This is a classic presentation of polycystic ovary syndrome where the patient struggles with significant hormonal changes, feeling quite low in mood, acne and significant deterioration to the quality of life and skin. The period was not on time.

The secret for success in TP's management is the fact that TP continued to comply with diet, exercise and the medication. TP will continue the current therapy for another two years before we consider withdrawing the treatment to see how well she will do without any further intervention. TP should continue to do well and there is no cause of concern.

SUCCESS STORY 5

Patient background

AM is a delightful 27-year-old nurse who was diagnosed with polycystic ovary syndrome in 2007 when she was 17 years of age. She had cysts in the ovary, she found it difficult to lose weight and her blood tests did not show any specific abnormality and she requested an expert second opinion. Her GP referred her to us for further assessment.

Referral to the endocrine unit

We examined AM in detail and we have noted that AM has not really been started on a specific treatment for the management of polycystic ovary syndrome.

When we met AM, she was concerned about infertility, cancer of the womb, weight gain, hair loss on the scalp and hair growth on the facial area. AM was joined by her mom and dad who also showed significant concern. I have agreed with AM that the increased luteinizing hormone to follicle stimulating hormone ratio is the cause of the excessive hair growth. The fact that her period is not on time and that she already has cysts on the ovary is important. Therefore, the diagnosis of polycystic ovary syndrome was not in doubt as AM's condition meets the Rotterdam criteria.

We explained to AM how important polycystic ovary syndrome is and how critical it is that she needs to address her weight carefully. AM was only 4 feet 9 inch and her weight was 9 stone 13 lbs. We agreed to try and reduce the weight to 8 and ultimately to 7.5 stone.

We agreed to an exercise program that will help AM during her busy schedule as a school nurse. We have discussed Spironolactone, Metformin and oral contraceptive pills as she is in a relationship and we have agreed to start the treatment in a specific order.

We explored why Spironolactone and Metformin are important. AM had a raised luteinizing hormone of 5.9 compared to a follicle stimulating hormone of 2.9 and that needs to be addressed.

Finally, we agreed on a low carbohydrate diet and on changing the exercise program that she currently does to ideally 30 minutes of exercise daily. We agreed that AM could not have the Spironolactone without having the oral contraceptive to make sure she does not get pregnant on Spironolactone. We have also agreed to check her Vitamin D and her Vitamin B12 to ensure that they are normal.

We met AM again with her family in the Clinic and we confirmed that the diagnosis of polycystic ovary syndrome is not in doubt, there is clear evidence of irregularity of the period, hair loss on the scalp and increased facial hair growth, difficulty losing weight, history of cyst in the ovary and she meets the Rotterdam criteria of 2003. AM by that time, managed to start on Spironolactone, Loestrin (oral contraceptive) and Metformin. She felt much better, her hair loss improved, and her weight improved. We agreed to increase the dose of medication.

Turning point

By 2017, AM felt so much better. Her energy level improved as well as the quality of the hair and the acne. Her weight improved from 9 stone 13 lbs. down to 9 stone. AM lost 13 pounds in weight by 2017. She is also aiming to lose further weight and the hormone profile continues to indicate a raised luteinizing hormone compared to the follicle stimulating hormone. We agreed to optimize AM's Vitamin D.

We reviewed AM again in 2018 and there has been further improvement in AM's wellbeing. The weight stabilized further from 9 stone 13 lbs. down to 9 stone, the period is on time, AM felt so much better and had more energy. AM was happy with the result of the hair and the quality of the skin and the mood has also improved. We agreed again that our target is to maintain the weight loss.

The plan is to wean AM off Spironolactone in the future and just to keep her on Metformin. AM's Vitamin D improved as well as her iron and her tests are regularly assessed to ensure she is safe to continue treatment.

Reflection

As you can see therefore, patients can respond to the treatment, but the treatment must be targeted and agreed on carefully by the patient. polycystic ovary syndrome is a very important condition. It cannot just be treated by saying go away and have an exercise program. That is not the key. The key area in managing polycystic ovary syndrome is diet. A low carbohydrate diet is the most critical part and the exercise compliments that. The medications only help a part of that. Also, it is important note that for sexually active patients who are on Spironolactone treatment, oral contraception is a must to avoid getting pregnant. Getting pregnant while on Spironolactone therapy can affect the way that the baby develops.

Metformin

Metformin is another important treatment for polycystic ovary syndrome which has been proven in many studies. Metformin is critical to help reduce the insulin resistance which is an issue in polycystic ovary syndrome. It has been advocated to prolong life, prevent type 2 diabetes and reduce the insulin level, and that helps reduce the impact of insulin on your weight.

Metformin is not confirmed as a sole weight loss drug however patients who are on Metformin notice significant improvement in their weight. The dose of Metformin is simple as 500 mg to 1000 mg per day and it should be around the evening meal time and that can help enormously in managing your insulin resistance overnight.

SUCCESS STORY 6

Patient background

CT is delightful 29-year-old physiotherapist who presented to us 3 years ago with difficulty losing weight and she was already referred to the National Health Service for help with weight management. She had been struggling with low mood, weight gain and lethargy. She had developed underactive thyroid and has strongly positive thyroid antibodies that attacked the thyroid gland. CT is very tired, fed up and exhausted.

CT had a personal trainer, exercised regularly at the gym and despite that she had been struggling to lose the weight. She was concerned that her thyroid may be an issue as she was on Thyroxine therapy since the age of 11.

CT's period started age 14. She developed problems with her period and the skin and she was started on oral contraceptive pills to help with that. However, she developed migraine despite trying different types of oral contraceptive pills and she could not tolerate Dianette and Minipill.

CT has been on and off the oral contraceptive pill for some time and found it difficult to use them. The diagnosis of polycystic ovary syndrome could not have been made on grounds of irregularity of the period since the period, of course, has been regulated using the oral contraceptive pill. The acne was mainly around the mandibular area (jaw line) and the upper back.

CT had evidence of hair growth on the abdominal wall and both above and below the umbilical area. In addition to that she had hair growth on the chest, the inner and back thighs, legs and toes.

What is quite clear is that CT had been struggling with increasing weight, excessive hair growth and was initially 10 stone in weight and that increased to 15 and a half stone. CT felt very physically tired after lunch and had to have caffeine to keep her awake. She was also diagnosed with irritable bowel syndrome sometime prior to that.

Therefore, we have a 29-year-old struggling with increasing weight, fluid retention, lethargy, mood change, presenting with skin changes at age of 14 and not able to tolerate the oral contraceptive pill with evidence of excessive hair growth in a number of areas and she meets the diagnostic Rotterdam criteria for polycystic ovary syndrome. The ultrasound is no longer required as we know that only two out of the three criteria are required to make a diagnosis of polycystic ovary syndrome.

We arranged a series of assessments and blood tests for hat revealed a low Vitamin D. We discussed the diagnosis of polycystic ovary syndrome with CT and acknowledged the need to address the Vitamin D and Vitamin B12.

Turning point

CT was started on Spironolactone and Metformin. She showed and felt a significant improvement in her wellbeing. Her weight dropped from 15 stone 7 lbs down to 13 stone 13 lbs. within one year. She has also noticed further improvement in the acne, weight, energy level and hair growth. We targeted a weight of about 10 stone.

We reviewed CT 6 months later and we were very impressed with how well she progressed. For example, the issue with the skin has already been resolved as well as the excessive lethargy and the hair growth. The headache has resolved, the fluid retention, tiredness, continuous bleeding throughout the month and the quality of the hair and the skin have all improved greatly.

A year later, CT continued to improve, and the weight dropped further down to 11 stone 11 lbs. and this was remarkable. This poor lady was carrying on four stone of excess weight and now felt so much better.

Once we took the polycystic ovary syndrome under control, it became clear that CT needed further help with her thyroid. We agreed that CT needed to start on Liothyronine and Thyroxine combination to optimise her thyroid function further.

This is a wonderful lady who had struggled for a significant amount of time at a young age for 29 years old with all the signs and symptoms of polycystic ovary syndrome. She was never diagnosed before at age 29 and with the diagnosis, came in the need to address her insulin level and her fluid retention, testosterone drive which had caused her most of the symptoms.

The next important part was to address her thyroid function. Thyroid problems develop in up to 40% of the patients with polycystic ovary syndrome and that also needs to be addressed. Vitamin D and Vitamin B12 must be optimised as Vitamin D deficiency affects hair and the skin while Vitamin B12 deficiency can cause significant lethargy.

Therefore, if you have polycystic ovary syndrome, please make sure that you have these specific areas addressed: your thyroid function, Vitamin D and B12 and your iron as these will impact upon your energy and ability to function.

Spironolactone

Another medication that has been advocated by the American Endocrine Society is Spironolactone. Spironolactone works by reducing the testosterone-like effect of polycystic ovary syndrome on you. It reduces excessive hair growth around the abdomen, the chest, the back of the legs or the toes and the acne. It also helps patients who have hair loss on the scalp. However, there are other products available from your local pharmacy such as hair, skin and nails, special organic based hair shampoo and hair treatments that can help as well in polycystic ovary syndrome patients who struggle with hair loss.

The dose for Spironolactone is 50 to 100 mg once a day and you must avoid potassium-rich diet. Your endocrinologist or your physician will be able to advise you further regarding its use. But, we know from the published data and the guidelines that Spironolactone is effective and safe so long as you hydrate well and avoid potassium-rich food such as bananas.

It has been advocated that surgery can help patients with polycystic ovary syndrome, but we have not seen any evidence of that. In the old days, we used to drain the ovary if the patient had an issue with conception.

SUCCESS STORY 7

Patient background

AG is a delightful lady who was 17 years of age when she was referred to us in 2016. Her period had stopped for three months and a pregnancy test was negative. She always had a regular period and she was on the oral contraceptive pill, Mercilon. The GP had documented a raised prolactin hormone of 548 and no other abnormality and asked for help.

Referral to the endocrine unit

We reviewed AG at the endocrine clinic and listened very carefully to AG's Mum and her stepdad. We agreed that there are a number of issues here that need to be addressed.

AG had her period stopped because she had polycystic ovary syndrome. We documented a blood test indicating clearly a raised luteinizing hormone compared to the follicle stimulating hormone. Her luteinizing hormone was 33, her follicle stimulating hormone was 7, she had skin changes and she had a raised prolactin because of polycystic ovary syndrome on her.

AG started her period age 11 and that was never on time. She developed skin changes on the forehead and the front of the chest. She tried oral contraceptive pills on two occasions and on both her period became regular.

We explained to AG the diagnosis of polycystic ovary syndrome and the impact of that on her. We have agreed that we must confirm that there is no evidence of pregnancy. We agreed that the acne that had developed on the forehead and the chest is part of the condition. AG was very slim and that is true for 50% of patients who have polycystic ovary syndrome.

However, in AG's case we were not satisfied, and we required further assessments. We organized for AG an ultrasound of the ovary which, as one expected, did not show any cyst since AG was on oral contraceptive pill for a while now. An MRI of the pituitary gland again did not show any abnormality as one would normally expect since the issue here was not a pituitary tumour causing excessive prolactin rather it was the polycystic ovary syndrome that is causing raised Prolactin.

Turning point

So why did we make the diagnosis of PCOs in AG's case?

We made the diagnosis of polycystic ovary syndrome on the ground that she had a disturbed hormone profile with a raised luteinizing hormone of 33 compared to follicle stimulating hormone of 7. There was evidence of acne and irregularity of the period in fact the period had stopped altogether and of course the cyst in the ovary was not required for diagnosis. This evidence confirms that there is an issue with the hormone profile.

We have explained to AG and Mum the diagnosis and we have agreed that AG will not need Metformin, she is already very slim, and she only needs Spironolactone and we have agreed on a specific type of oral contraceptive pill for her.

AG was reviewed lately in the clinic and she looks so much better. She is full of energy, excellent quality of the hair and the skin and her period was on time and she does not have any other issues at the time.

Reflection

50% of patients with polycystic ovary syndrome can be very slim and 50% of patients with polycystic ovary syndrome can put on weight. Not all patients with polycystic ovary syndrome have excessive hair, acne and not cysts in the ovary. Patients may have a raised prolactin for many reasons e.g. pregnancy, stress, pituitary tumour or medication. However, in AG's case, the raised prolactin was due to polycystic ovary syndrome. Every patient presents with a unique picture that can be any variation of that.

Therefore, when you have a referral to your endocrinologist, you must be aware that you may not have the full picture of polycystic ovary syndrome to present as a patient who needs a treatment for polycystic ovary syndrome. It is important however that you make sure that if you are on oral contraceptive pill to explain clearly to your endocrinologist that you have been on it. If you have acne and you have history of irregularity of the period, then there is a high chance that you have polycystic ovary syndrome even you are slim.

SUCCESS STORY 8

Patient background

RP is a delightful young lady. We look after RP's Mum who has underactive thyroid and a significantly low vitamin D. RP's Mum raised concerns about her daughter and asked us to review her. We listened very carefully to RP who explained that she has had a degree of anxiety, tiredness and struggled with a tightness across the chest. She felt quite low in mood, she struggled with excessive hair growth in the abdomen, the toes, the legs and the chest and the acne over the face, the shoulder and the chest.

We agreed to assess RP for several issues such as polycystic ovary syndrome, vitamin B12 deficiency, thyroid problem, vitamin D deficiency and we have agreed with RP to have a series of investigation and then sit with her as soon as these investigations are ready to review her progress.

We reviewed RP a few weeks later and it is very clear that RP is struggling with polycystic ovary syndrome. The evidence are clear, she has difficulty with anxiety and tiredness, she has mood change, acne, excessive hair growth and sweats affecting her quality of life. Her vitamin D was only 47 (normal level should be between 75 to 150), and there is a family history of thyroid cyst. She had clear evidence of raised luteinizing hormone to follicular stimulating hormone ratio. We agreed that RP has polycystic ovary syndrome. We also agreed that the ultrasound is not necessarily required for diagnosis as RP already met two out of the three Rotterdam criteria. We started RP on treatment with Metformin and Spironolactone and she adjusted her diet and agreed on a specific exercise program.

Turning point

We reviewed RP a few months later and we could see a significant improvement in her quality of life. Poor RP struggled with significant mood swings. She noticed an improvement in many areas on the oral contraceptive pill. When RP stopped the oral contraceptive pill for a break, she noticed a recurrence of the symptoms and we agreed that RP will try a specific oral contraceptive pill which will help her manage the hormone profile.

We looked again at RP in more detail a few months later and we noticed further improvement. Her luteinizing hormone to follicular stimulating hormone has improved. We agreed to continue the Metformin, the Spironolactone, the oral contraceptive pill, especially as now she is in a relationship and she cannot continue the Spironolactone without the oral contraceptive. We agreed to start RP on Citalopram, which is an antidepressant to help improve her mood.

We discussed family support to RP and we agreed to a specific way of trying to help RP adjust to her current work commitments. We agreed also that RP has a number of areas to address with her expectations in terms of her work commitment. That is something that RP needed to discuss with her line manager to help her adjust her work load.

Reflection

A 21-year-old, therefore, struggling with mood swings and excessive sweating, acne, hirsutism and a raised luteinizing hormone to follicular stimulating hormone confirming a diagnosis of polycystic ovary syndrome that is affecting her quality of life. What really brought RP to the clinic is not the symptoms of excessive hair growth but the mood changes and the acne.

Often ladies do cope well with the excessive hair growth but cannot cope very well with the mood changes that tend to happen in polycystic ovary syndrome. Patients with polycystic ovary syndrome often struggle with anxiety. It becomes a problem when the patient's symptoms upset her to a degree that she is not able to function anymore.

It is not unusual that we refer a patient to the clinical psychologist to support them with that. It is important to realise the impact of the disturbance in testosterone and insulin levels on the mind and the ability to cope with anxiety. It is critical that you are able to address that in addition to the physical symptoms since the psychological part of polycystic ovary syndrome is equally as important.

Victoza

Another important drug that has recently come to be scientifically considered for the management of polycystic ovary syndrome is a drug called Victoza (GLP-1 analog) which is an important drug that we use currently in the management of type 2 diabetes patients. Victoza is now started to be used in management of patients with polycystic ovary syndrome in a number of studies and it is a drug that has a major impact on the weight.

In a recent publication in the Lancet; a reputable medical journal in May 2017, it was indicated clearly that 80% of patients who are prediabetic (having a slightly increased blood sugar but not yet diabetic) and struggling with their weight, lose a significant amount of weight using Victoza.

It has been published in many journals how effective Victoza is in trying to help you manage the weight better. Victoza is usually given at night as one injection before you go to bed.

How does this drug work?

Victoza works by managing the insulin resistance and it has a major impact on satiety (feeling full). When a patient looks at food, there is a stimulant in the brain that triggers an electric activity to the gut. The gut then produces a hormone called Glucagon-type Peptide 1 and that then goes to the pancreas and adjusts the insulin level, goes to liver and reduces the amount of sugar coming out, it goes to the brain and induce the feeling of fullness and slows down the gut motion.

Victoza is an "analog" or a similar molecule to Glucagon-type peptide and therefore works in a similar way to reduce the blood sugar, promote satiety and reduce body weight. Victoza brings down the blood sugar to a normal level without causing an excessively low blood sugar (hypoglycaemia). You may develop some nausea when you start Victoza as a side effect. The good news about this drug is that it is given only by injection once at night and has a significant impact on the patient weight.

Victoza is used specifically in patient with type 2 diabetes and there is new research indicating how effective it is in patients with polycystic ovary syndrome and that was published by a Dutch group in 2014 and there has been significant success of Victoza in patients with polycystic ovary syndrome.

SUCCESS STORY 9

Patient complaint

SS was referred to the endocrine clinic in 2009 when she was 19 years old. Her body mass index was 32 (healthy body mass index should be 18.5 to <25), with a weight of 80 kg. She found it difficult to lose weight since 2006, three years prior to being referred. Both her Mum and SS had excellent diet.

SS was reviewed by the local endocrine team for management of her underactive thyroid. SS tried medications to help her lose weight such as Orlistat (a medication that minimises fat absorption from the diet), however she did not really get any benefit. Both SS and her Mum were convinced that there is a reason for the difficulty losing weight despite the diet and exercise and having a personal trainer and that is why they asked for an expert second opinion.

Referral to the endocrine unit

We reviewed SS at the endocrine clinic and we assessed her condition in more detail. A diagnosis of polycystic ovary syndrome was made on the fact that her period was not regular, and she was started on the oral contraceptive pill age 16. SS discontinued the oral contraceptive pill at age 19 and her period then became regular simply because she has been on the oral contraceptive pill for three years. She has difficulty losing weight and she has noticed excessive hair growth (hirsutism). We agreed with SS that the target weight to aim for is 9 stone 8 lbs. (61 kg) while her current weight at the time was 12 stone 9 lbs. (80 kg).

Turning point

SS is now on Metformin and Spironolactone. She noticed significant improvement in her wellbeing. We then referred SS to the gynecology team for further assessment as SS developed per-vaginal bleeding. SS continued to attend the clinic regularly and her condition continued to improve, and she felt so much better. She found it difficult sometimes to control the diet. However, her weight improved from her original weight of 80 kg down to 73 kg. SS continued the Thyroxine, Metformin and Spironolactone and continued to do well.

So, the secret for success for SS is the fact that she adopted the diet and exercise with meticulous hard work and continued to improve because of her effort to control the diet very well. It has not been easy for SS to control the polycystic ovary syndrome however she managed to do so very well.

In 2016 SS luteinizing hormone and follicle stimulating hormone had improved significantly and we have agreed that the best way forward is to control her polycystic ovary syndrome with the oral contraceptive pill, Metformin and Spironolactone.

Her thyroid function was managed on Thyroxine and Liothyronine (T3) as SS's thyroid function was not controlled on Thyroxine alone. SS continued to attend the clinic on a regular basis. She currently is regularly reviewed by the endocrine unit and she is now age 28.

SS battled significantly with her weight. Her weight continued to fluctuate, and we agreed that the priority for SS is to be comfortable and happy with her current weight. SS is managed to control the significant problem with the skin and the hair.

We met again in early 2018 and we discussed the weight in more detail and we have discussed why SS is struggling with such a problem with the weight and we have agreed to consider Victoza which is a drug that is based on hormones specifically to help the weight, the blood sugar and the insulin level.

SS has struggled with managing her thyroid. She is very fit, very active and she has tried her best with diet and exercise and despite Metformin and despite Spironolactone, she continued to struggle with the degree of insulin resistance that caused weight gain and that is why we've decided to consider Victoza for SS and we expect SS to improve and control her weight using that.

At one stage in 2016 we discussed referring SS for liposuction surgery and we have agreed that it is not the best way forward for a young person and it will be best to manage the polycystic ovary syndrome with diet and exercise, Metformin, Spironolactone and now two years down the line we are lucky enough to be able to use Victoza as a drug that would help PCOS patients lose their weight.

Common questions

Other questions that patients often ask are:

How long do I stay on Metformin? How long do I stay on Spironolactone? Whether for the rest of my life?

The answer is not as simple and clear as that. What we know is that Metformin is helpful in managing insulin resistance and insulin resistance is a condition that lasts all your life if you have polycystic ovary syndrome. It is not a condition that disappears because you have developed the menopause. It is a condition that must be managed all your life and the best way to manage it is diet, exercise and Metformin.

Spironolactone is slightly different. Spironolactone is best used at the time when you are really struggling with excessive hair growth, acne or hair loss of the scalp and normally patients will be on it for one to two years and then they come off it when they have had very good control over the acne or the hair growth and if the symptoms recur again they can restart it again.

It is important however to note that you cannot be on Spironolactone unless you have very good contraceptive methods such as oral contraceptive pills or the Mirena coil. You must not get pregnant while you are on Spironolactone as it can harm the baby. It is important that you do not become pregnant if you are on Spironolactone and if are trying to become pregnant you must not be on Spironolactone for a three months' time before trying.

SUCCESS STORY 10

Patient complaint

CH is a 49-year-old lady who has had hormonal issues for many years. She has a history of multiple miscarriages and heavy regular periods. CH had a son when she was age 26 and she had fertility issues and needed fertility assistance earlier when she was 31 years old to try and conceive and was successful to deliver a baby boy who is now 16 years of age. She presented to her GP with the complaint of excessive hair loss that was upsetting her. She has had the Mirena coil about one year ago which improved her heavy periods to a degree. The GP referred her to the Endocrine Clinic for assessment of her hormone profile and to help with her complaint.

Referral to the endocrine unit

CH attended to the Endocrine clinic in January 2018. We reviewed CH and listened to her carefully. She was upset by the degree of hair thinning and hair loss she was experiencing. She complained of hot flushes that disturbed her sleep and was not coping well. CH helped us keep track of a series of hormone tests she has had since 2000 which was very helpful.

CH tests showed a disrupted luteinizing hormone to follicle stimulating hormone ratio. So now, we have a delightful lady who has a history of fertility issues, she has heavy periods, hair thinning and loss and history of cysts in the ovary in addition to abnormal luteinizing hormone to follicle stimulating hormone ratio.

The diagnosis of polycystic ovary syndrome was previously made in London and was not in question. CH had a series of careful investigations that showed a low level of Vitamin D, Vitamin B12, iron and declining estradiol hormone. CH was initially started on hormone replacement therapy to compensate for the decline in estradiol. She was also started on Vitamin D supplementation which is very critical to improve the hair loss CH is struggling with. Iron and Vitamin B 12 supplements were also started to try and bring their level up to a better stand to help CH with her wellbeing and health.

Turning point

We are watching CH carefully through regular blood tests to monitor her safety and response to treatment. It is very important to try and normalise the low oestradiol level which is why CH is on Hormone Replacement Therapy Oestradiol patches at the moment.

Low oestradiol level is responsible for the hot flushes and contributes greatly to the hair loss too. Spironolactone will then be considered once we assess CH for response to the oestradiol treatment. We must note that it is very important to keep an eye on patient's safety first and not to overload CH with medications that might cause her unpleasant side effects. This is the reason why we take treatment in a stepwise approach, resolving issues one at a time and monitoring the response closely.

We are keeping an eye on CH's thyroid function. As we know that thyroid dysfunction tends to occur in 20 % - 40% of patients with polycystic ovary syndrome and it needs to be assessed regularly. CH is not diabetic which is a good sign of her insulin level and this is also being checked carefully to make sure we catch any abnormality in CH's blood sugar early on.

Reflection

This case reveals the fact that polycystic ovary syndrome does not simply disappear as ladies approach the menopausal transition. It may in fact be complicated and worsened by the decrease in oestradiol level which adds to the hair loss which is a very common symptom of this condition. It is also very important not to burden a patient with too many treatments. Rather, the best approach is to slowly introduce treatments to correct abnormalities one issue at a time.

Careful assessment of possible complications of the condition is very important to make sure we track the patient's health with a keen eye instead of just waiting around for another problem to develop. If you have a diagnosis of polycystic ovary syndrome, make sure you have a careful assessment of your ferritin, Vitamin D, Vitamin B12, oestradiol, cortisol and thyroid functions regularly.

SECTION 5

PREGNANCY

Pregnancy and polycystic ovary syndrome

Pregnancy for a patient with polycystic ovary syndrome often causes a concern and worry to young patients. The means by which you can manage and improve the outcomes of their pregnancy is very critical. Most of our patients have questions about this when they first attend to the endocrine clinic.

Do you think my polycystic ovary syndrome will affect my fertility?

This is often a problem and polycystic ovary syndrome, yes can affect your fertility if you are not treated. However, patients with polycystic ovary syndrome when well-treated, should not have any issues with fertility.

Infertility develops when the polycystic ovary syndrome is active, when the hormone profile is not well controlled, when the weight increases or when there is a disturbance to the prolactin or the thyroid or the blood sugar or there is clear evidence of the irregularity of the period.

If you control the polycystic ovary syndrome well with diet, exercise and Metformin then you should not have any issues with fertility and if you do then there are other areas that you need to explore as I will explain here.

How can I get pregnant when I have the diagnosis of polycystic ovary syndrome?

It is important to try and make sure that you have addressed the diet in detail, have set up a clear exercise program that I have explained above. Also, you need to check that you are seeing your partner often enough at least twice a week to make sure that you are together twice a week rather than seeing each other once a month. That is not helpful in trying to get pregnant in any patient whether you have polycystic ovary syndrome or not.

In a study performed in Italian couples found that they can improve fertility in 50% of patients who have presented with infertility issues by advising them simple measures. One of them is to try to see their partners often enough but not rarely as the mere fact of being together can improve your fertility.

If you are struggling to get pregnant it is important to make sure that you have been already on Metformin for at least 3 to 6 months and then you need to check your prolactin, thyroid function and the blood sugar.

You need to continue to try and improve your health and increase the chances of success of pregnancy by simply having a regular rest in the afternoon, managing your stress level, as stress may increase the level of cortisol (the stress hormone) which can in itself raise your insulin level further and cause you further problems with managing your polycystic ovary syndrome.

It is important that you address your weight and ensure that you are within a healthy body mass index, measure your waist circumference and check that you have a normal waist circumference and check your blood sugar and ensure that your sugar level is within normal range.

It is also important to check that you have a normal level of Vitamin D, check that you are not anaemic and check that your thyroid function is normal. If you are struggling to become pregnant even though you have already adjusted the diet, the exercise, lost weight and ensured that you have a normal thyroid function, prolactin level and blood sugar then it is important that you check your thyroid antibody.

If your thyroid antibody is positive, then your physician may consider targeting your thyroid stimulating hormone level to be within a specific range according to the local guidelines to try and make sure that your thyroid is optimum for you to try and become pregnant.

If all fails, then you can consider a course of Clomid, which is normally a tablet once a day for five days per month to try and help you induce ovulation and that is often successful in combination with Metformin.

If all fails then a last resort is to consult with an IVF expert to check that there are no other reasons to explain why you are not getting pregnant. In our clinic in the last 15 years the number of patients who have been referred to IVF was hardly 10. Simply because the patients have been successful by the clear management plan above which is diet, exercise, Metformin, ensuring adequate and normal

Vitamin D, iron level and ensuring that the thyroid, prolactin and the blood sugar are all within normal limits.

Therefore, in summary, to become pregnant if you have polycystic ovary syndrome you need to follow the normal usual best practice rule of managing polycystic ovary syndrome. That means ensuring that your weight is normal, and you address your weight by starting on a diet and exercising as highlighted above. You should have a clear idea what your thyroid, prolactin, blood sugar levels are and your thyroid antibody status. Consult with your physician regarding the possibility of Clomid. If all fails, then IVF therapy is the last possible option.

Please try and simplify the treatment of polycystic ovary syndrome. It is not a condition that is not treatable, although it is a condition that is not curable however you can improve it significantly.

SUCCESS STORY 11

Patient complaint

OJ is a delightful 30-year-old lady beautician was referred to us 5 years ago with an increasing headache, muscle cramps, and starting on Liothyronine (T3 underactive thyroid therapy). She was under the care of a previous endocrinologist. Her main symptoms were mind fogginess and poor concentration. She had started on the T3 therapy by the endocrinologist as she had a low T3 levels. Despite the general practitioner increasing her T3, she did not improve and hence referred her to the unit.

Referral to the endocrine unit

When we first met OJ who was a beautician five years ago, her main complaint was the poor quality of hair. She was very weepy, very upset and she lost her mum sadly to meningitis at a young age. She saw the gynecologist at the local NHS hospital. She was concerned about the symptoms of polycystic ovary syndrome especially with the acne and she has a problem with the symptoms of premenstrual tension, mood swings, low mood, hot flushes and she felt quite unwell and tired with low abdominal discomfort and bloating. Her cycle was not regular, and she suffered from severe menstrual pain.

OJ was started on oral contraceptive pill and was referred to the endocrinologist who organised for her further assessment. The endocrinologist noted that she does not have any other specific abnormality other than a low T3 and she was then started on T3 therapy. She had an ultrasound which indicated that she had polycystic ovaries.

The symptoms were very clear, she had thinning of the hair on the scalp, brain fogginess, anxiety and stress. OJ period started age 15 and she was started on the other contraceptive pill at age 15 because of acne and heavy periods and stopped the oral contraceptive pill at 29 prior to the presentation at 30 years of age. That is why then her symptoms of polycystic ovary syndrome appeared. Specifically, the oral contraceptive pill was helpful for her and she doesn't struggle with excessive hair growth but just does struggle with acne.

We discussed with OJ the importance of insulin resistance and the impact of that on her, the importance of the impact of testosterone on her and have discussed with OJ how important it is to ensure that she has a treatment for polycystic ovary syndrome as soon as possible to try and improve her fertility. She was started on Metformin and Spironolactone with very good effect. We have agreed that OJ will continue the Dianette and address the vitamin D.

Turning point

Nine months later, OJ was reviewed in the clinic and she felt so much better. Her mood and quality of the hair and the skin have improved, and she does not suffer from ovarian pain. We agreed that weight loss and controlling the sugar craving were the next. We have agreed to a specific exercise program and diet.

This was critical since OJ is an excellent cook and a very good baker and she always insists on bringing some of the cakes to the clinic and it's very difficult for us to refuse. We agreed that OJ would need to focus on managing the polycystic ovary syndrome and there is no indication for her to continue the Liothyronine.

If we look carefully at OJ two years later, we find that OJ continues to improve. For example, her energy level, hair, skin, and fluid retention have significantly improved. The pelvic pain has improved and OJ adjusted the diet, increased the Metformin and optimised her Vitamin D. OJ was not on Thyroxine anymore as she did not need treatment for her thyroid, rather she needed treatment for the polycystic ovary syndrome.

We reviewed OJ again three years down the line, and there has been further improvement and we organised further assessments. We increased the dose of Metformin and continued the current dose of Spironolactone. There has been an improvement in her insulin resistance and she felt much better. OJ was then reviewed on a number of occasions the following year.

Two years ago we discussed further the practicality in terms of managing the fertility and we reassured OJ that she does not have any reason not to start a family. OJ planned to start a family in 2016. We adjusted the treatment accordingly and discontinued Spironolactone which must be discontinued at least three months prior to trying to conceive. A year later, we reviewed OJ with an excellent result and she was 29 weeks pregnant. OJ now has a lovely healthy boy with no cause of concern.

Reflection

This case illustrates the importance of an accurate diagnosis in patients with polycystic ovary syndrome. The symptoms of underactive thyroid are very easily confused with the symptoms of polycystic ovary syndrome. The symptoms can be similar as patients tend to put on weight, feel tired, show hair loss on the scalp. It is important to address the symptoms correctly. It is commonly misunderstood that the thyroid is an issue when a patient presents with symptoms polycystic ovary syndrome.

It is then frustrating for the patient, the consultant and the GP to find that the thyroid function test is normal. Only by persisting and addressing the actual clinical scenario that the patient feels better. This is an example of how the patient feels much better having addressed the main issue which is the polycystic ovary syndrome.

What do you do when you become pregnant and you have polycystic ovary syndrome?

This is a common question that a lot of patients ask, and the answer is very straight forward, and the treatment is exactly as you would do for polycystic ovary syndrome patients before pregnancy. This is to ensure that you have a regular clear dietary plan with regular intake of vegetables and proteins with reduced amount of carbohydrates and fats in your diet. You must have regular check-ups with your midwife or gynecologist or your primary care physician. You must avoid at all cost type 2 diabetes which can develop in polycystic ovary syndrome patients who are pregnant and the way to do that is to follow a low carbohydrate diet.

Please explain to your medical team about polycystic ovary syndrome as not many people would be aware of this important condition and its impact upon you. It is important that you watch your blood pressure and know what your blood pressure is and to record it carefully and help your physician check it for you on regular basis.

Ensure that you have a clear plan for breast feeding and ensure that you discuss that issue with your midwife. If you are struggling to consider breast feeding you can discuss the benefits of breast feeding with your obstetrics team. If you are worried about breast feeding, then there are measures that the team can help with to help you improve and address your concerns.

If you are worried that you have excessive urination or that you are thirsty, it is important to highlight that quickly to your team to check your blood sugar and watch the last few weeks of pregnancy as they are the most difficult. You need to be careful and have regular rest and ensure that you have good communication with your obstetrics and maternity team to help you through this important phase of your pregnancy especially the last 12 weeks of pregnancy.

Some physicians prefer to keep you on Metformin during your pregnancy, some physicians do not, it depends on the local guidelines.

SECTION 6

COMPLICATIONS

Patient complaint

RE was referred by her GP to be assessed for increased sleep disturbance due to having to wake up at night to pass urine. We have tried her on a number of medication and they thought that they might have a problem with her pituitary gland causing a condition called Diabetes Insipidus. RE was then referred to the endocrine clinic for further assessment of these concerns.

Referral to the endocrine unit

We assessed RE in more detail at the endocrine clinic and it was causing significant concern to note that RE had more than just what could be a pituitary issue.

RE had a very complex medical condition. She was age 56 when she presented. She had a cyst on the ovary removed at age 24, another cyst at age 40, she had a hysterectomy at age 40, HRT at age 50 and by age 56 she was burdened by a significant difficulty of losing weight and fluid retention to a degree where she had to wake up at night every 90minutes to open the bladder. The diagnosis was a late diagnosis of polycystic ovary syndrome causing her significant insulin resistance and unstable bladder that developed following having a hysterectomy to remove the womb.

She has had a history of prediabetes and had a condition with significant cysts developing under the armpit and in the groin. She developed a condition called sleep apnea where she found it difficult to sleep properly at night due to obstruction of the airways. She ended up having a stroke and having significant headaches and she had two surgeries to remove ovarian cysts, but she never had a diagnosis of polycystic ovary syndrome.

The reason she developed all these problems is the lack of treatment of polycystic ovary syndrome. The polycystic ovary syndrome caused RE to develop significant insulin resistance which then caused the diabetes. She had cysts in the ovaries and abnormal hormonal profile resulting in cysts under the armpit and in the groin. The increase in fluid retention caused the problem with the weight. She was waking up at night every 90 minutes to open her bladder simply because she was developing a weak bladder. Although RE had surgery on two occasions for removal of cysts in the ovaries, she never really had a treatment for polycystic ovary syndrome.

Turning point

We started RE on diet and exercise, Metformin and Spironolactone. She did remarkably well and lost a total of 11 pounds within the first six weeks. She is on target to lose a total of three stone to try and prevent any further complications of the polycystic ovary syndrome on her.

Reflection

As you can see from this important case, it is important to try and control the polycystic ovary syndrome before the patient develops the menopause.

It is important to make sure that with the treatment of the polycystic ovary syndrome the patient is being protected from developing any further problems such as diabetes, high blood pressure, and stroke. The risks for these complications is inherently increased by development of menopause.

This also highlights the fact that patients with polycystic ovary syndrome do not lose the polycystic ovary syndrome symptoms and its complications having developed the menopause. In fact, after the menopause, it is critical to control the polycystic ovary syndrome more efficiently to try and avoid any problems with the development of stroke or diabetes or high blood pressure or major problems with fluid retention.

The key point to point out here is polycystic ovary syndrome is not a condition that is resolved by surgery or cured by excision of the cysts. polycystic ovary syndrome is a systemic condition that affects the entire body in different aspects. Failing to understand this aspect of the condition will lead to major complications that can severely the patient's quality of life and health.

What other conditions can be caused by polycystic ovary syndrome?

The three conditions that polycystic ovary syndrome can often be associated with are:

- Raised prolactin
- Disturbance to the thyroid gland
- Disturbance to the blood sugar causing diabetes

Raised Prolactin

Patients with polycystic ovary syndrome tend to have a raised prolactin which is hormone produced from the pituitary gland. The pituitary gland is a gland based in the middle of the brain that controls our hormone system and a raised prolactin can often disturb or stop the period and can cause breast discharge. Testing for this can be very simple through a blood test that you can ask your physician to help you with.

Treatment for a raised prolactin is extremely simple with a tablet taken once a week which can reduce the prolactin level back to normal. However, it is important not to just start on the treatment but also to have a clear assessment of why your prolactin is raised. Could that be simply due to polycystic ovary syndrome or other causes such as a problem with your pituitary gland. Any patient with a raised prolactin level is often referred to the endocrinologist for further assessment and if you have polycystic ovary syndrome that is a common cause. Other causes such as the use of oral contraceptive pill, stress or a pituitary issue can also cause a raised prolactin.

Hypothyroidism

Patients with polycystic ovary syndrome also tend to have hypothyroidism in a significant proportion of patients and that can contribute to weight gain and that is why it is important also to have your thyroid checked.

Type 2 Diabetes

In terms of type 2 diabetes, there is a significant risk of developing type 2 diabetes if you have polycystic ovary syndrome. However, you can prevent it aggressively with managing your diet and exercise and ensuring that you address the polycystic ovary syndrome in a very constructive consistent way every day by managing the diet and exercise as the main target for you to manage the polycystic ovary syndrome. By managing the polycystic ovary syndrome with diet and exercise your risk of developing diabetes diminish significantly especially if you are on Metformin.

Polycystic ovary syndrome and Gynecological issues

Polycystic ovary syndrome can cause gynecological problems such as irregularity of the period, the period can stop or be very painful, or development of a significant number of cysts that can cause acute abdominal pain around the pelvic area.

It is always an advice that if you have polycystic ovary syndrome that you must have a gynecology review to ensure that you have no issues from a gynecology point-of-view. We would normally recommend that you see a gynecologist once a year to make sure that the pelvis is healthy with no issues of concern.

It is important that you have 8 periods per year to avoid the development of abnormal cells within the womb which can turn into cancerous cells. Therefore, if you don't have 8 periods every year you must have that checked by your gynecologist and you have regular examinations of the womb.

My mood and polycystic ovary syndrome

We know that patients who have polycystic ovary syndrome often feel upset and angry, change in mood very quickly and they feel that they just cannot control their moods. This can be associated with a disturbance in their hormone level at the time of the period or the week before their period. This can also be associated with disturbance to their blood sugar and it can be easily managed by managing their blood sugar level carefully throughout the day.

If you have any mood issues or you are struggling or feeling angry or feeling low in mood, it will be important to discuss that with your primary care physician and it is also important to consider treatment for polycystic ovary syndrome with more than just diet and exercise. We know that patients on Spironolactone help stabilise their moods better and some patients will need an antidepressant for a short period as polycystic ovary syndrome can disturb the serotonin level.

Sleep disturbance and polycystic ovary syndrome

Patients often complain of waking up at 3 am in the morning and then finding it difficult to get back to sleep or having disturbed sleep or have nightmares. This can usually be associated with disturbance to their sugar level prior to their sleep or with the raised cortisol level that they have developed because of the polycystic ovary syndrome. The stress of the day upon them can raise cortisol level enough to disturb their sleep.

The reason you wake up in the morning is that your cortisol level rises enough to wake you up and the reason that you go back to bed at night is your cortisol level drops at night to help you go back to bed and sleep. This is called a circadian rhythm, and this is what controls us in term of the clock of our body to help us sleep and rest at night and wake up fresh in the morning. Therefore if you have disturbed sleep, your cortisol level will rise and that can cause your health problems such as patient who work on night shifts develop diabetes, they develop difficulty to lose weight and high blood pressure and that is why it is important to have a good sleep pattern to go back to bed at a regular time and try to sleep for good 7 to 8 hours every night.

If your sugar level is disturbed at night because you of course are asleep and you are not eating and the insulin level is high, that can drop your blood sugar in the middle of the night typically around 3 to 4 o'clock in the morning, waking you up and you find it difficult to sleep and therefore I commonly advise that you have something to eat before you go to bed. Eat something simple like a bowl of fruits before you go to bed so that it can help you for the next 8 hours when you are asleep and not eating and your blood sugar could be dropping and waking you up.

I'm so tired

The best way to improve your energy level if you have polycystic ovary syndrome is to ensure that you have an adequate sleep pattern and to make sure that your thyroid function is regularly checked. Make sure that you have a regular exercise program and that you have a clear dietary plan. Also, you must make sure that you have exercise for a specific amount of time throughout the week. You need to write down the amount of exercise you have done per day and calculate it on a Sunday and find out if you have achieved 350 minutes of regular exercise. Remember this is the amount of exercise that is required to improve your insulin resistance and it is important that you document clearly that you have done exercise every day.

It is also important to make sure that you have adequate Vitamin D intake and that your level of Vitamin D has been measured and is in the normal range according to the WHO criteria and American Endocrine Society guidelines. It is also important to know that if you are on Metformin, you should be on Vitamin B12 and if you haven't had your Vitamin B12 checked, you should ask your GP to help you to check your Vitamin B12.

It is important also to know that if your thyroid function test is normal today, that does not necessarily mean that it will be normal in a year or two. Thyroid can fail at any time, the thyroid can become overactive and then suddenly underactive and that is a condition called viral thyroiditis and it may help you to read Volume I in the series of "My Thyroid and Me" which will help you understand the variation in the thyroid function and how best to look after your thyroid.

Another way to try to improve your energy level is to try and restrict Gluten in your diet, reduce the fat in your diet, increase the amount of protein in your diet and reduce carbohydrates in your diet. Simple measures such as these: good exercise, good diet and good sleep pattern, should result in improvement in your energy level. If that does not help, then you should consult with your primary care physician and your endocrinologist to explore why you are tired. Other conditions that could then be explored could be obstructive sleep apnea or disturbance to your cortisol level which require careful assessment by your endocrinologist.

SECTION 6

FINAL WORDS

Continue to learn how to look after your self

The last few questions

Several patients ask about polycystic ovary syndrome and whether it is curable?

We confirm that polycystic ovary syndrome is treatable but not curable. To date we do not have enough data to suggest how best to manage this genetically inherited condition. It is well known that you can control it very well by simply controlling the polycystic ovary syndrome with diet, exercise, Metformin and considering Spironolactone if indicated.

How do I explain polycystic ovary syndrome to my family?

The best way to explain polycystic ovary syndrome to your family is to explain that it is a condition that causes you high insulin level and that requires treatment with diet and exercise and sometimes medication.

The future for polycystic ovary syndrome is bright there has been a significant explosion of data to suggest that there are ways to try and help polycystic ovary syndrome patients further. There is a very good group in Massachusetts General Hospital and in Boston who have been working very hard to try and help define the best treatment option for polycystic ovary syndrome. If any further data develop I will include it in the next version of this book.

What to remember?

Manage stress in a very positive way, try and find a system that works for you to manage your stress. Stress is part of your daily life whether it is personal life or work life and you must make sure that you have a system to manage the stress.

Please avoid sleep deprivation. If you are sleep deprived, it tends to increase your insulin level and your cortisol level and that can cause you significant problems with managing your polycystic ovary syndrome and you will end up becoming quite unwell. Sleep deprivation is the worst torture you can do to your body, so you must make sure that you have an appropriate sleep pattern. I have a very dear patient Wendy, who is a very busy executive and she has always taught me that the best way to manage your sleep is to have a regular 8-hour sleep per day.

Plan your diet in a weekly schedule, do just go for what is available, plan a week ahead or a month ahead. Weigh yourself once a week rather than once a day, adjust your diet accordingly to result in a response. Don't just wait till the weight drops off, you need to have a feedback and don't worry if you are not losing weight in the first few weeks, simply because there is a significant insulin resistance that needs to be addressed before you start to lose weight and the evidence that you are improving the insulin resistance is the weight loss. The first sign of weight loss is your waist circumference.

Learn how to improve your health. It is important that you become an expert in improving your health. Study carefully what I have said in this book to help you manage your diet and exercise, manage your sleep pattern, manage your thyroid, manage your cortisol level, manage your Vitamin D, manage your iron and Vitamin B12 to make sure that you have regular diet, exercise and sleep pattern. Limit the portion of your food and choose your food carefully. Have two days break where you have only vegetables and salads to rest your bowel and insulin and to allow your body to recover from all that.

Remember there are so many diets out there. This is not a diet book, this is a manual to help you manage the polycystic ovary syndrome with the diet in mind. However, you need to manage the polycystic ovary syndrome.

Use this book as a manual

Keep this book at home or in your car and consult with it for a few minutes every day to learn something new about polycystic ovary syndrome or read it over coffee. This book is not to be read once but it is a manual to help you manage the polycystic ovary syndrome which is a lifetime condition. You should always try and consult with this book all the time to try and help you on regular basis how best to manage this important condition.

Please make notes in yellow sticky paper to mark specific sections of the book or if you download it on Kindle simply make notes in your own diary of certain parts of the book that you should read again. This is a complex medical condition which I have tried to simplify as much as possible.

Please read it and learn it every day for a few minutes rather than spend hours reading it. Simply because it is a complex condition and it requires your careful attention. Ask the question and find the answer in the book. If you cannot find the answer in the book, please let us know and we will help you as much as we can.

*"Every great dream begins with a dreamer.
Always remember, you have within you the strength, the patience, and the passion to reach for the stars to change the world".*

Harriet Tubman

Printed in Great Britain
by Amazon

77175581R00047